YOUR KNOWLEDGE HAS VALUE

- We will publish your bachelor's and master's thesis, essays and papers

- Your own eBook and book - sold worldwide in all relevant shops

- Earn money with each sale

Upload your text at www.GRIN.com and publish for free

Line Following with Obstacle Avoidance using BOE-BOT

Bandar Hezam

Bibliographic information published by the German National Library:

The German National Library lists this publication in the National Bibliography; detailed bibliographic data are available on the Internet at http://dnb.dnb.de.

ISBN: 9783346980045
This book is also available as an ebook.

Print and binding: Books on Demand GmbH, Norderstedt, Germany
Printed on acid-free paper from responsible sources.

The present work has been carefully prepared. Nevertheless, authors and publishers do not incur liability for the correctness of information, notes, links and advice as well as any printing errors.

GRIN web shop: https://www.grin.com/document/1426587

Intermediate Robotics
Individual Assignment

TITLE	**Line Following with Obstacle Avoidance using BOE BOT**
NAME	Bandar Naji Ali Hezam
SUBMISSION DATE	2 /8/19

Acknowledgement

In preparation of my report, I had to take the help and guidance of some respected persons, who deserve my deepest gratitude. As the completion of this report gave me much pleasure, I would like to show my gratitude to Dr. ARUN SEERALAN BALAKRISHNAN, Course Instructor ITR, on APU University for giving me a good guideline for report throughout numerous consultations. I would also like to expand my gratitude to all those who have directly and indirectly guided me in writing this Assignment.

In addition, a thank you to Mr. NAZRI HADI, who introduced me to the Methodology of work, and whose passion for the "underlying structures" had lasting effect. Many people, especially my classmates have made valuable comment suggestions on my paper which gave me an inspiration to improve the quality of the report. I would also like to express my gratitude to APU's technicians of the laboratory for giving me the resources needed to complete this assignment.

Abstract

In this assignment, programing the BOE-BOTS for maze solving was required. The BOE-BOTS were tasked to follow a specific line from the start to the end. In addition to that, there would be obstacles that the BOE-BOTS need to face. The BOE-BOTS will have to avoid those obstacles and reach the end. This report was written for the purpose of discussing the process of developing the program used in this assignment for the maze solving. Firstly, it will provide the reader with a technical background so they can understand the process of development of the program made for this assignment. Secondly, the used codes would be keenly analyzed and explained. Third and lastly, the challenges that was encountered in the process of developing and the process of developing itself will be talked about in detail followed by a conclusion of the whole report.

Table of Contents

Table of figure

Introduction

Robotics is the industry related to the engineering, an expansive and differing field identified with numerous business enterprises and buyer employments. The field of mechanical autonomy by and large includes taking a gander at how any physical developed innovation system can play out an undertaking or assume a job in any interface or creative innovation. It manages structuring, creating and working a robot utilizing programming and sensors to execute explicit assignments. These robots are utilized to supplant people and perform errands all the more effectively. It likewise used to perform complex assignment which are past the span of human capacity. As innovation progresses, assignments are getting to be simpler these days where portable robots are more created than any time in recent memory. In this task, a portable robot must be customized and complete 2 undertakings. A versatile robot is a computerized robot, which has the capacity of headway (Habib, 2015). Portable robots can move unreservedly around a situation relying upon what they are modified for. A portable robot can perform activities like shipping loads starting with one spot then onto the next, investigate the earth or other complex errands utilizing locally available frameworks. These locally available frameworks incorporate servomotors, which turn the wheels, mechanical arms, sensors and some more. For this task a BOE BOT will be utilized and customized to finish certain errands. A BOE-BOT portable robot is worked with Arduino microcontroller, servomotors and an Arduino shield. The BOE BOT will be customized utilizing Arduino code and transferred onto the Arduino microcontroller. Arduino is an open-source stage utilized for creating gadgets advancements. Arduino comprises of both a physical programmable circuit board, which is frequently alluded as a microcontroller, and a bit of programming, or an IDE (Integrated Development Environment) that keeps running on the PC which is utilized to compose and transfer PC codes to the physical board. The Arduino stage ended up prominent with the headway of innovation where a great many people began utilizing gadgets. In contrast to other programmable sheets, the Arduino does not require a different bit of equipment to stack a refreshed code into the board where a USB link can be utilized to transfer the refreshed variant of the code. Also, the Arduino IDE utilizes an improved rendition of C++ where it makes it simpler to program and get the hang of programming. At last, Arduino gives a standard structure factor that breaks out the elements of the miniaturized scale controller into an increasingly available bundle. The BOE-BOT should achieve undertakings like exploring itself through the line following labyrinth and recognizing any deterrent on its way. To finish these kinds of errands, a BOE-BOT utilizes a mix of sensors and actuators to detect nature and make a particular move until the destinations needed are fulfilled. A sensor is a gadget that can distinguish a sort contribution from the physical condition and can react to these kinds of information sources. There were two sensors utilized in the BOE-BOT which are the Infrared sensor and the Ultrasonic sensor. Every one of these sensors were utilized for a particular errand to achieve the principle target which is following the line and maintaining a strategic distance from hindrances. An infrared sensor is an electronic instrument that is utilized to detect certain qualities of its environment. It does this by either producing or identifying infrared radiation. Infrared sensors are likewise equipped for estimating the warmth being transmitted by an article and recognizing movement. (Chilton, Oct 15, 2014) Infrared Spectroscopy's hypothesis has been around since F.W. Herschel found infrared light in 1800. Herschel made an examination utilizing a crystal to refract daylight and had the option to identify the infrared radiation past the red piece of the noticeable range utilizing a thermometer to quantify an expansion in temperature. Be that as it may, the IR sensor was utilized in the BOE-BOT to identify the lines and react contingent upon the program made. The IR sensor comes in pair where it comprises of an IR recipient and an IR producer. The producer is a light radiating diode which emanates infrared light and an IR collector is a photodiode which is delicate to infrared light discharged by the IR producer. For this situation, the IR sensor will distinguish the differentiation of the line and the encompassing where it will recognize high contrast and send a comparing sign to the Arduino microcontroller.

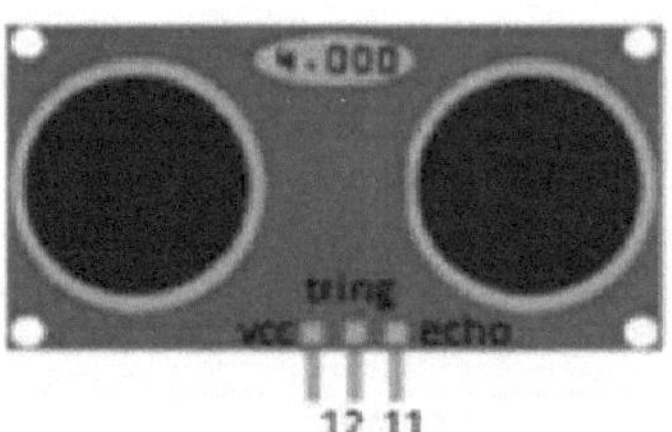

Figure 1:Ultrasonic

The Ultrasonic sensor is the second sensor that was utilized in the BOE-BOT. Ultrasonic sensor is a sort of sensor that is intended for non-contact estimation. This sensor utilizes a solitary ultrasonic component for both emanation and gathering. The Ultrasonic sensor utilizes the ultrasonic waves to quantify the separation of an article. An optical sensor which has a transmitter and recipient is called ultrasonic sensor. It discharges the waves which is thought about the article and got back by the sensor. It at that point estimates the separation by ascertaining the time taken between the discharge of the waves and getting it.

Figure 2:BOE-BOT

Line follower robots have played a significant role in industries growth. Initially, used for the purpose of delivering mails within an office building, they are now serving to deliver medication in a hospital. With more accuracy in the design, the technology could even be suggested for running buses and other mass transit systems, and may end up as part of autonomous cars navigating the freeway. Many automation techniques help to design such robots.

Objective

The main objectives of this assignment are:

1- To program a mobile robot to navigate through line maze with the ability to avoid obstacles in its path way.

2-To gain extra knowledge in areas such as sensors, programming and control system.

Programs for line following and Obstacle Avoidance

So as to make the BOE-BOT explore through line labyrinth easily and stay away from the impediments effectively, it ought to be appropriately customized. This program was made in the Arduino programming. The Arduino programming utilizes a programming language, which is only a lot of C/C++ capacities that can be called from the code, which is the point at which the library capacities are used for controlling the outside equipment, for instance. sensor and servomotors. In the first guide, there was just one begin way and two END paths additionally this line labyrinth is included a fixed well in center between two mobile wells. The BOE-BOT is to begin from the START spot and returns once more from the end ways to the begin path By utilizing ultrasonic sensor, the line supporter can distinguish an impediment and can stop till the hindrance is evacuated by as appeared in Figure 2. The program was likewise made to dodge and return back when the Ultrasonic estimates the separation to an article by utilizing sound waves. It will quantify the separation by conveying a sound wave at a specific recurrence and listening that wave when it skips back. The Map that the BOE-BOT ought to pursue is appeared in Figure 2. From the figure, we can see that the guide comprises of various bends and snags. The BOE-BOT ought to have the option to go from the earliest starting point, following the blue bolts that I added to the figure, stay away from the deterrent that is appeared in the guide and completes toward the END. From the guide, we can see that there are intersections which the BOE-BOT needs to cross more than one time. In this way, every intersection should be tried to ensure that it very well may be modified to take the most straightforward course. This is to ensure that pointless extra time won't be spent. In this manner, the line adherent sensors ought to be coded such that will guarantee that when the intersection is distinguished, it needs to make an alternate move. To play out this, a counter was inputted into the program.

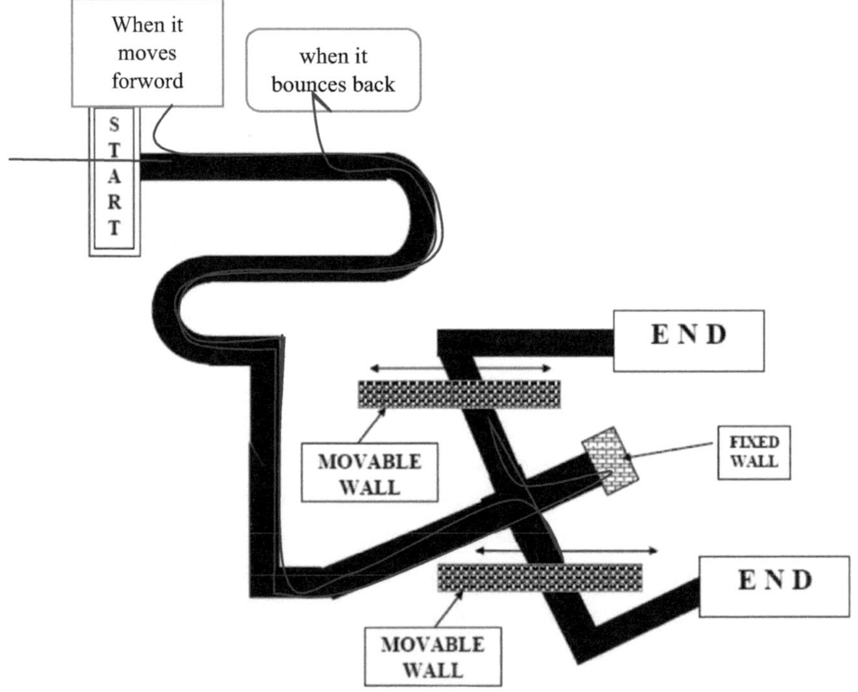

Figure 3:LINE MAB

The Arduino codes begin with certain comments where the primary comment expresses the utilization of the program and after that it gives some depiction about the software engineer.

Some data about the program is additionally added to make a simpler comprehension of the program. Right off the bat, the servo library was incorporated into the program. This is indispensable where the servo library permits an Arduino board to control the servo engines. From that point forward, the servo engines, which are the left servo engine and the correct servo engine, are introduced. At that point, a steady whole number has been characterized for the Trigger stick and the reverberation stick. These pins are the pins which are associated from the ultrasonic sensor to the Arduino. At that point, two factors are proclaimed, which are span and separation, with the goal that it very well may be utilized for programming the Ultrasonic sensor. from that point onward, a few factors which are b1, a2, n3, d4, and r5 are announced as whole numbers. These factors speak to the line supporter sensor inputs. Every factor speaks to every sensor of the line devotee sensors. They are instated as indicated by the info stick number that is associated in the Arduino. Two factors are likewise pronounced. These two factors are named as counter1 and counter. They are instated with an estimation of 0. Counters were utilized to perform various undertakings under a similar condition

```
// this program is used for maze solving
// Programmed by: Saher Samir TP043253
// Counters have been added when one conditions has two commands.
// Counters are initialized with a name of counter_2 and counter_3.
// Time delay was used to smoothen the line following
#include <Servo.h>
  Servo servoLeft;
  Servo servoRight;
  int a1 = 4;
  int a2 = 5;
  int a3 = 6;
  int a4 = 7;
  int a5 = 8;
 int counter_2=0;
 int counter_3=0;
// // defines pins numbers
const int trigPin = 9;
const int echoPin = 10;
// defines variables
long duration;
int distance;
```

Figure 4:Declaring Variables

The program begins with certain comments where the main comment expresses the utilization of the program and after that it gives some portrayal about the program. Some data about the program are likewise added to make a simpler comprehension to the program. Right off the bat, the servo library was incorporated into the program. This is imperative where the servo library permits an Arduino board to control the servo engines. Servos have coordinated apparatuses and a pole that can be absolutely controlled. This library likewise contains a few distinct capacities inside it to control the servomotors in various ways. A portion of the capacities in this library incorporate "compose Microseconds, append (), compose ()". Anyway just "compose Microseconds" and "append ()" will be utilized to control the servos. From that point forward, the servo engines, which are the left servo engine and the correct servo engine, are introduced. At that point, a few factors which are a1, a2, a3, a4 and a5 are pronounced as numbers. These factors speak to the IR sensor inputs. Every factor speaks to every sensor of the IR sensors. They are introduced by the info stick number that it associated in the Arduino. Two factors are likewise proclaimed. These two factors are named as counter_2 and counter_3. They are instated with an estimation of 0. Counters are fundamental where by utilizing counters, various undertakings can be performed in a similar point. From that point forward, a steady whole number has been characterized for the Trigger stick and the reverberation stick. These pins are the pins which are associated from the IR sensor to the Arduino. At that point, two factors are announced, which are span and separation, with the goal that it very well may be utilized for programming the Ultrasonic sensor

```
void setup()
{
  pinMode(trigPin, OUTPUT); // Sets the trigPin as an Output
pinMode(echoPin, INPUT); // Sets the echoPin as an Input
  Serial.begin(9600);
  pinMode(a1,INPUT);
  pinMode(a2,INPUT);
  pinMode(a3,INPUT);
  pinMode(a4,INPUT);
  pinMode(a5,INPUT);

  servoLeft.attach(13);
  servoRight.attach(12);
//Moving forward
  servoLeft.writeMicroseconds(1600);
  servoRight.writeMicroseconds(1400);

}
```

Figure 5:Setup function

After delaring the variables needed and including the libraries needed, the Setup function was utilized. This capacity would just be executed once. This capacity is utilized to introduce the factors and setting the pins for the info and the output. As it very well may be seen from Figure 3, pinMode was utilized. PinMode designs the predetermined stick to act either as an info or a yield. At first, the trigger stick was set as a output where the reverberation stick was set as an information. Reverberation stick and Trigger stick are utilized to speak to the Ultrasonic sensor. SerialBegin(9600) is for the transmission of information between the PC and the program.

From that point onward, the five factors, which are a1, a2, a3, a4 and a5, are set as an info. These factors are utilized to speak to the IR sensor. As there are 5 sensors, every factor speaks to one sensor. From that point onward, the capacity join, which is taken from the library, is utilized to characterize every servo engine and the stick that it is associated. In this code, the left servo engine is associated in stick 13 where the correct servo engine is associated in stick 12. From that point onward, the BOE-BOT was modified to move straight. This was finished by causing the left servo engine to go counter clockwise, which is characterized as 1600, and the correct servo engine will turn clockwise, which is characterized as 1400. Be that as it may, it will push ahead gradually. This was finished by adjusting the wheels to stop at 1500.

```
void loop()
{
  int s1 = digitalRead(a1);
  int s2 = digitalRead(a2);
  int s3 = digitalRead(a3);
  int s4 = digitalRead(a4);
  int s5 = digitalRead(a5);
// Clears the trigPin
digitalWrite(trigPin, LOW);
delayMicroseconds(2);
// Sets the trigPin on HIGH state for 10 micro seconds
digitalWrite(trigPin, HIGH);
delayMicroseconds(7);
digitalWrite(trigPin, LOW);
// Reads the echoPin, returns the sound wave travel time in microseconds
duration = pulseIn(echoPin, HIGH);
// Calculating the distance
distance= duration*0.034/2;
// Prints the distance on the Serial Monitor
Serial.print("Distance: ");
Serial.println(distance);
```

Figure 6:Loop function

After the setup function, the loop function is used. This function makes the codes repeat throughout the process. Five integers are initialized where these variables equal to the output of the digital read function. The digital read function reads the value from a specified digital pin, either HIGH or LOW. In this code, the pin is a1-a5, which are input pins of the IR sensor. After that, the digital write function was used. The digital write functions write a HIGH or LOW value to a digital pin. This function was used to write the trigger pin as LOW for two microseconds to clear it. Then, the same function was used to write the trigger pin as HIGH for ten microseconds. Subsequently, it is written back to LOW by the same function. Then, the integer duration is defined where it is equal to the output of the function pulseIn. PulseIn function is used to read the pulse of the pin whether it is HIGH or LOW. It returns the pulse in microseconds or return to zero. The pin that was defined was the echo pin. Then, the distance variable is initialized by an equation as seen in Figure 4. After that, a function, which is serial.print, is used to print the distance in the serial monitor. This was necessary to check whether the sensor is working properly or not where the Serial monitor can be opened, and the sensor can be tested. Figure 4 shows the codes that were written in the beginning of the loop to set and initialize some variables which are related to the sensor.

```
if (distance<19)
{
//Moving slightly right
  servoLeft.writeMicroseconds(1650);
    servoRight.writeMicroseconds(1650);
delay(250);
// Moving forward slowly
  servoLeft.writeMicroseconds(1600);
    servoRight.writeMicroseconds(1400);
    delay(400);
}
else{

  if (s1==0 && s2==0 && s3==0 && s4==0 && s5==0)
  {
    // Moving back to adjust
  servoLeft.writeMicroseconds(1450);
    servoRight.writeMicroseconds(1650);
    delay(150);
  }
    else if (s1==0 && s2==0 && s3==1 && s4==0 && s5==1)
  {
    // Moving Slightly right to adjust
  servoLeft.writeMicroseconds(1550);
    servoRight.writeMicroseconds(1500);
delay(150);
  }
```

Figure 7: Main program

After initializing and setting some variables that related to the sensors, the actual coding that makes the BOE-BOT follow the line and avoid the obstacles are programmed. Firstly, the code that makes the BOE-BOT avoid the obstacle is being programmed. This was done by making an if statement which indicates that if the distance, which is a variable that was declared, is less than 15, the BOE-BOT had certain commands to follow. These commands are to Move slightly to the right for 5 seconds. Then, move forwards slowly for 5 seconds. The variable distance indicates the distance from the BOE-BOT to the obstacle. These commands will only be executed only if the distance is less than 15. The distance is being indicated by the Ultrasonic sensor which was used to sense any object that is in-front of the sensor. The commands that made the BOE-BOT go right or left was programmed using the write. Microseconds function. This function has an input which is an integer. This integer was assigned depending on servo whether if the programmer wanted it to move clockwise or counter-clockwise where 1500 was calibrated as the stopping of the two wheels. Then the counter 1 value is increased by 1 to be used in the other cases.

Figure 6 also shows the codes that were programmed if there is no obstacles in-front of the sensor. Inside the else statement of the first If statement, another if statement was made. These if statements are used to make the BOE-BOT follow the line. The second if statement's condition states that if s1, s2, s3, s4 and s5 are equal to zero, the left servo motor should go clockwise, and the right servo would go counter-clockwise. This would make the whole BOE-BOT move backwards. This was done where if the BOE-BOT went out of the line, this command will make the BOE-BOT adjust itself back to the line. The variables s1 to s5 represents the IR sensors where 0 indicates that the sensor does not read any black line and 1 indicates that the sensor is reading a black line. It may be noticed that a delay function was also used. The delay function makes the commands before it execute for a certain time where this time is in microseconds. In this code, the delay function had an input of 150 microseconds which means 1.5 seconds. After the if statement, an else if statement was added where its condition was if s1, s2 and s4 equal to 0 and s3 and s4 equal to 1, the BOE-BOT should turn slightly to the right. This was used to adjust the robot back to the line. The commands to go slightly right was programmed by making the right servo motor stop rotating where the input was 1500, which was calibrated to stop. In the other hand, the left servo motor would rotate counter-clockwise which would make the BOE-BOT go slightly right. The time delay function was added to smoothen the process of the line following.

```
   else if (s1==1 && s2==1 && s3==1 && s4==1&& s5==1)
  {
     // making an if statement for the counter
 if (counter_3 ==0)
 {
  // when it senses it, the first time, it will go infront slowly
       servoLeft.writeMicroseconds(1550);
   servoRight.writeMicroseconds(1450);
    delay(150);
    // then, it will go slightly right
       servoLeft.writeMicroseconds(1550);
   servoRight.writeMicroseconds(1550);
    delay(800);
    // then, move infront
          servoLeft.writeMicroseconds(1550);
   servoRight.writeMicroseconds(1450);
    delay(150);
    // counter value will increase by 1
 counter_3 =counter_3 +1;
   }
 else if(counter_3 ==1)
  // when it senses it, the second time, it will go right slowly
 {
   servoLeft.writeMicroseconds(1550);
    servoRight.writeMicroseconds(1550);
   delay(250);
   // then, it will go forward
   servoLeft.writeMicroseconds(1550);
   servoRight.writeMicroseconds(1450);
   delay(300);
   // counter value will increase by 1
   counter_3 =counter_3 +1;
   }
```

Figure 8: Counter

There were different else if statements that were added to the program. However, these codes have the same concept. This means that there were other conditions that were added in the else if statements. These conditions were assigned accordingly. The difference between the explained codes and the other codes are either the condition of the code or the commands. The condition of the code differs where there are other conditions that the sensors senses. These conditions were given specific commands according to the sensor's conditions. The commands were added in the same way where the servo is being defined first. Then, the writeMicroseconds function is added with an input. This input depends on the movement of the BOE-BOT that is required. For example, if the sensor senses 1, 1, 0, 0, 0, there would be commands for the BOE-BOT to go slightly left. These codes to perform this task would be the same as the previous codes. However, the input of the writeMicroseconds function will change. Even though all the if statements are having the same concept, there was an else if statement that had some changes in it. These changes are in terms of the commands that are placed inside the else if statement. These commands were different because a counter has been added to the program. A counter has been added been to the program so that when the sensor senses the same alignment but each time the sensor senses one alignment, a specific task needed to be executed. It was vital to assign a counter because there was an intersection in the map where in the first time, the BOE-BOT needed to turn right where the second time it needed to take a different route. For example, Figure 8 shows the usage of a counter. In figure 8, all the 5 sensors sense the same thing which is the black line. In the map, the BOE-BOT passes twice in the same point. However, the task is changed in each of them. Therefore, to assign each point a task, an if statement has been made. The condition of this statement was if that counter is equal to zero, which is initialized in the beginning of the code, the following commands are executed. When the counter has the value of zero, it means that this is the first time that the BOE-BOT passes from this point. The commands in this statement are to move in-front slowly for 1.5 seconds. Then, it is to go right for 8 seconds. After that, it is to straight for 1.5 seconds. Finally, the counter is initialized a new value which is the previous value added with 1. This was coded so that in the next intersection code, the value of counter will change, and the following codes will be performed in the else if statement. The second time the sensor will sense all ones, the code will only read the else if statement where the condition states that the counter_3 is equal to 1. The following codes will be executed. It will first go right slowly with a time delay of 2.5 seconds. Then, it will go forward for 3 seconds. After that, the counter's value will increase by 1. Similarly, one more case where a counter was added was implemented in the program. The process of developing the counter was the same. However, the commands to either go right, left, straight or backwards change accordingly.

Discussion

The task was finished effectively where two sensors were connected to the BOE-BOT for finishing the labyrinth and maintaining a strategic distance from the deterrents. The principle destinations of the task were accomplished. This can be said where two unique sensors were connected to the BOE-BOT which are the Infrared Sensor and the Ultrasonic sensor. The infrared sensor is utilized to identify the white and dark line and send a sign to the board and respond as indicated by what it is customized. Correspondingly, the ultrasonic sensor is utilized to identify the snags and send a sign to the board where it gives an order to keep away from it. The board that was utilized was the Arduino where it could program every one of the sensors and servomotors by transferring a code from the Arduino programming to the board. In any case, there were a few challenges that was confronted while programming the labyrinth comprehending codes.

These troubles confronted made numerous imperfections in the program made. The primary issue looked during this task was because of the sensors joined to the BOE-BOT. Having two unique kinds of sensors implies that every sensor needs a power supply. For this situation, the power supply for these sensors was a battery. There was where if the battery was not charged adequately, the IR sensor would have a power cuts which caused an issue following the line. This would likewise make a postponement in the IR sensor as far as sending and getting signals. Also, there were a few issues confronted while making the BOE-BOT pursue the line easily. This is a result of the distinctive BOE-BOTS that were utilized while playing out this task. For instance, one of the BOE-BOT had a period deferral of 100 milliseconds where it accomplished a smooth immaculate turning. In any case, a similar code with a similar time postponement was transferred to another BOE-BOT, it didn't turn easily and there was some abnormality while it was turning. By investigating these variables and difficulties, there were some sharp perceptions made. Right off the bat, the sensors must be cleaned where it must be ensured that there is no residue on them. Since even the scarcest residue can influence the manner in which the sensor works, and in this way impacts the way that the BOE-BOT works. Also, the wires are a significant angle where it must be associated appropriately. On the off chance that there are any wires which are free or not associated appropriately, it will prompt inadequacy of the BOE-BOT. At last, it was seen that the scarcest residue particles on the guide impacts the way that the BOE-BOT works. This is might cause delays. This is on the grounds that the residue particles may influence the manner in which that the IR sensor works. Be that as it may, the difficulties confronted were defeated where the BOE-BOT has effectively pursued the line and maintained a strategic distance from the hindrances.

Conclusion

As an end, the BOE-BOT was customized. It was modified utilizing Arduino controller. The sensors and actuators were associated with the controller. The control of the movement is allotted utilizing Arduino programming. The code has been determined by the way set. The learning results are accomplished, and tasteful. Different kinds of sensors for mechanical application, for example, line following, and obstruction shirking is used. The program results are accomplished just as the capacity to distinguish building issues and apply designing standards to take care of the issues are picked up. Ultrasonic sensor was utilized to moveable item on the guide. The infrared sensor was utilized to identify the lines in labyrinth. The information on end effector was picked up when planning a fitting end effector for the robot. Be that as it may, some little time postponements were made by the Boe-Bot. As it was clarified before, they could be tackled as the battery has full charge. The residue, the little illustrations and earth were influencing the little blemish. In spite of having flaw in the robot, the task can be considered as effective. Along these lines, the goals of this task are accomplished.

References

M. Sri Venkata Sai Surya1, K. B. R. K. P. K. S. S. M., April 2018. *Smart and Intelligent Line Follower Robot with Obstacle Detection* , Kattankulathur. Chennai – 603203. India: 4Department of Electrical and Electronics Engineering, SRM Institute of Science and Technology,.

Appendix

```
// this program is used for maze solving
// Programmed by: Saher Samir TP043253
// Counters have been added when one conditions has two commands.
// Counters are initialized with a name of counter_2 and counter_3.
// Time delay was used to smoothen the line following
#include <Servo.h>
  Servo servoLeft;
  Servo servoRight;
  int a1 = 4;
  int a2 = 5;
  int a3 = 6;
  int a4 = 7;
  int a5 = 8;
 int counter_2=0;
 int counter_3=0;
//  // defines pins numbers
const int trigPin = 9;
const int echoPin = 10;
// defines variables
long duration;
int distance;
void setup()
{
  pinMode(trigPin, OUTPUT); // Sets the trigPin as an Output
pinMode(echoPin, INPUT); // Sets the echoPin as an Input
  Serial.begin(9600);
  pinMode(a1,INPUT);
  pinMode(a2,INPUT);
  pinMode(a3,INPUT);
  pinMode(a4,INPUT);
  pinMode(a5,INPUT);
  servoLeft.attach(13);
```

```cpp
  servoRight.attach(12);
//Moving forward
  servoLeft.writeMicroseconds(1600);
  servoRight.writeMicroseconds(1400);
}
void loop()
{
 int s1 = digitalRead(a1);
 int s2 = digitalRead(a2);
 int s3 = digitalRead(a3);
 int s4 = digitalRead(a4);
 int s5 = digitalRead(a5);
 // Clears the trigPin
digitalWrite(trigPin, LOW);
delayMicroseconds(2);
// Sets the trigPin on HIGH state for 10 micro seconds
digitalWrite(trigPin, HIGH);
delayMicroseconds(7);
digitalWrite(trigPin, LOW);
// Reads the echoPin, returns the sound wave travel time in microseconds
duration = pulseIn(echoPin, HIGH);
// Calculating the distance
distance= duration*0.034/2;
// Prints the distance on the Serial Monitor
Serial.print("Distance: ");
Serial.println(distance);
if (distance<19)
{
//Moving slightly right
 servoLeft.writeMicroseconds(1650);
  servoRight.writeMicroseconds(1650);
delay(250);
// Moving forward slowly
 servoLeft.writeMicroseconds(1600);
```

```cpp
  servoRight.writeMicroseconds(1400);
  delay(400);
}
else{
 if (s1==0 && s2==0 && s3==0 && s4==0 && s5==0)
 {
 // Moving back to adjust
 servoLeft.writeMicroseconds(1450);
  servoRight.writeMicroseconds(1650);
  delay(150);
 }
   else if (s1==0 && s2==0 && s3==1 && s4==0 && s5==1)
 {
 // Moving Slightly right to adjust
 servoLeft.writeMicroseconds(1550);
  servoRight.writeMicroseconds(1500);
delay(150);
 }
/////
 else if (s1==0 && s2==0 && s3==0 && s4==1 && s5==0)
 {
   // Moving Slightly right to adjust
 servoLeft.writeMicroseconds(1550);
  servoRight.writeMicroseconds(1500);
  delay(150);
 }
   else if (s1==0 && s2==0 && s3==0 && s4==1 && s5==1)
 {
   // Moving Slightly right to adjust
 servoLeft.writeMicroseconds(1550);
  servoRight.writeMicroseconds(1500);
delay(150);
 }
 else if (s1==0 && s2==0 && s3==0 && s4==0 && s5==1)
```

```cpp
{
  // Moving Slightly right to adjust
servoLeft.writeMicroseconds(1550);
  servoRight.writeMicroseconds(1500);
delay(150);
}
  else if (s1==0 && s2==0 && s3==1 && s4==0 && s5==0)
{
// Moving forward slowly
servoLeft.writeMicroseconds(1600);
  servoRight.writeMicroseconds(1400);
delay(150);
}
  else if (s1==0 && s2==0 && s3==1 && s4==1 && s5==0)
{
  // Moving Slightly right to adjust
servoLeft.writeMicroseconds(1550);
  servoRight.writeMicroseconds(1500);
delay(150);
}
  else if (s1==0 && s2==0 && s3==0 && s4==1 && s5==0)
{
  // Moving Slightly right to adjust
servoLeft.writeMicroseconds(1550);
  servoRight.writeMicroseconds(1500);
delay(100);
}
  else if (s1==0 && s2==0 && s3==1 && s4==1 && s5==1)
{
// making an if statement for the counter
  if (counter_2==0)
  {
  // when it senses it, the first time, it will go infront slowly
servoLeft.writeMicroseconds(1600);
```

```cpp
servoRight.writeMicroseconds(1400);
 delay(300);
 // counter_2 value will increase by 1
counter_2 = counter_2 + 1;
 }
 else if (counter_2 ==1)
 {
  // when it senses the second time, it will go infront slowly
   servoLeft.writeMicroseconds(1600);
 servoRight.writeMicroseconds(1400);
 delay(300);
 // counter_2 value will increase by 1
counter_2 =counter_2 +1;
 }
 else if (counter_2 ==2)
 {
  // when it senses it the third time, it will go front slowly
   servoLeft.writeMicroseconds(1600);
 servoRight.writeMicroseconds(1400);
 delay(150);
 // then goes right slowly
   servoLeft.writeMicroseconds(1550);
 servoRight.writeMicroseconds(1550);
 delay(800);
 // then moves forward slowly
    servoLeft.writeMicroseconds(1550);
 servoRight.writeMicroseconds(1450);
 delay(150);
counter_2 =counter_2 +1;
 }
 else {
  // if it senses other than the defined conditions, it will go straight slowly
 servoLeft.writeMicroseconds(1600);
 servoRight.writeMicroseconds(1400);
```

```cpp
  delay(200);
    }
}
  else if (s1==0 && s2==1 && s3==0 && s4==0 && s5==0)
  {
  // moving left slowly
servoLeft.writeMicroseconds(1500);
  servoRight.writeMicroseconds(1400);
delay(150);
}
  else if (s1==0 && s2==1 && s3==1 && s4==0 && s5==0)
  {
  // moving left slowly
servoLeft.writeMicroseconds(1500);
  servoRight.writeMicroseconds(1400);
delay(100);
}
  else if (s1==1 && s2==0 && s3==0 && s4==0 && s5==0)
  {
  // moving left slowly
servoLeft.writeMicroseconds(1500);
  servoRight.writeMicroseconds(1400);
delay(100);
}
  else if (s1==1 && s2==1 && s3==0 && s4==0 && s5==0)
  {
servoLeft.writeMicroseconds(1500);
  servoRight.writeMicroseconds(1400);
delay(100);
}
  else if (s1==1 && s2==1 && s3==1 && s4==0 && s5==0)
  {
servoLeft.writeMicroseconds(1550);
  servoRight.writeMicroseconds(1450);
```

```cpp
delay(200);
}
  else if (s1==1 && s2==1 && s3==1 && s4==1&& s5==1)
  {
    // making an if statement for the counter
if (counter_3 ==0)
{
// when it senses it, the first time, it will go in front slowly
    servoLeft.writeMicroseconds(1550);
  servoRight.writeMicroseconds(1450);
  delay(150);
  // then, it will go slightly right
    servoLeft.writeMicroseconds(1550);
  servoRight.writeMicroseconds(1550);
  delay(800);
  // then, move infront
    servoLeft.writeMicroseconds(1550);
  servoRight.writeMicroseconds(1450);
  delay(150);
  // counter value will increase by 1
counter_3 =counter_3 +1;
  }
else if(counter_3 ==1)
// when it senses it, the second time, it will go right slowly
{
  servoLeft.writeMicroseconds(1550);
  servoRight.writeMicroseconds(1550);
  delay(350);
  // then, it will go forward
  servoLeft.writeMicroseconds(1550);
  servoRight.writeMicroseconds(1450);
  delay(300);
  // counter value will increase by 1
  counter_3 =counter_3 +1;
```

```cpp
  }
else {
    // if it senses other than the defined conditions, it will go straight slowly
    servoLeft.writeMicroseconds(1550);
  servoRight.writeMicroseconds(1450);
   delay(200);
    servoLeft.writeMicroseconds(1550);
  servoRight.writeMicroseconds(1550);
   delay(800);
     servoLeft.writeMicroseconds(1550);
  servoRight.writeMicroseconds(1450);
   delay(200);
  }
}
}
```

YOUR KNOWLEDGE HAS VALUE

- We will publish your bachelor's and
  master's thesis, essays and papers

- Your own eBook and book -
  sold worldwide in all relevant shops

- Earn money with each sale

Upload your text at www.GRIN.com
and publish for free